P9-ECO-910

ON THE LOOSE

SIERRA CLUB • BALLANTINE BOOKS

ON THE LOOSE

BY JERRY & RENNY RUSSELL

Illustrated with 63 brilliant photographs

As for small difficulties and worryings, prospects of sudden disaster, peril of life and limb; all these, and death itself, seem to him only sly, good-natured hits, and jolly punches in the side bestowed by an unseen and unaccountable old joker... There is nothing like the perils of the wilderness to breed this free and easy sort of genial, desperado philosophy. —MELVILLE

Fletcher once requested, in an interview, that he not be labelled an "adventurer." That word, he said, is associated with irrespon-sible individuals, and his feat of outdoorsmanship required fastidious detail work. —SAN FRANCISCO CHRONICLE

What is life but a series of inspired follies? —SHAW

The photographs in this book are of the lowest fidelity obtainable. They are as far from the photographer's vision as cheap cameras, mediocre film, and drugstore processing could make them.

The design is by Terry Russell

We are grateful to the following authors and publishers for quotations used within this book:

De la Mare, Walter. From *Song of Finis,* by permission of The Literary Trustees of Walter de la Mare and the Society of Authors as their representative.

De Voto, Bernard. From *Course of Empire,* by permission of Houghton Miflin Company.

Joyce, James. From *A Portrait of the Artist as a Young Man,* by permission of The Viking Press, Inc.

Leopold, Aldo. From *Sand County Almanac,* by permission of Oxford University Press.

Shaw, George Bernard. From *Pygmalion,* by permission of The Public Trustee and the Society of Authors.

McQueen, Steve. By permission of Steve McQueen.

Wells, H. G. From *Tono Bungay,* by permission of Dodd, Mead & Company.

Paperback Edition by Ballantine Books, New York
Library of Congress Catalog Card No. 67-18013

Manufactured in the United States of America

To Ma,
who worried.

When I was young, and reckless too, and I craved the reckless life...

—Folk Song

I am glad I shall never be young without wild country to be young in. Of what avail are forty freedoms without a blank spot on the map? —ALDO LEOPOLD

Have you ever walked 34 miles on a straight-arrow dirt road in the desert with only a Tang-jar of some rusty water because you expected somebody who didn't come and then walked past your turnoff in the dark and had to sleep on a cattleguard? Have you ever dropped your sleeping bag in the ocean by mistake? Have you ever followed a jeep-track in the rain which got worse and worse and fainter and fainter and petered out a vertical quarter mile from where you wanted to go? Have you ever slept on a cobblestone river bank? or on a sand dune on a windy night and spit sand all the next morning? Have you ever climbed a mountain but missed the right peak by half a mile but the sun was down and you were freezing and had better find some dry wood and a place to sleep in the snow quick? Have you ever walked 234 miles of mountain trail to see how fast you could do it? Have you ever started a backpack trip and hit a storm on the first pass and spent 24 hours under a wet plastic tarp drinking lumpy icy chocolate and walked through the snow to a cabin and burnt your jeans drying them over a wood stove? Have you ever left your insect repellent behind on a rock? Have you ever had a cheese sandwich for Christmas dinner in Death Valley? Have you ever camped in a dump? Have you ever gone to sleep on a beach and woke up in water and had to sleep up on rocks under a cliff which rained sand on your neck all night and lost a tennis shoe and almost your glasses to the tide? Have you ever lain under a truck for five hours because it was the only shade in the desert in July? Have you ever walked 50 miles? or walked 41.3 miles with blisters for glory? Have you ever fallen out of and under a boat in a rapid because the deck wasn't tied on right? Have you ever had just dried figs and sandy bread for breakfast, lunch, and dinner? Have you ever floated a lake shore at night groping for a campsite midst bare rock and cactus? Have you ever built a fire with a water ski because it was the only wood?

No, I reckon not all of them, maybe. But that's how we've grown up, Ren and I: that and a thousand little glimmers on the water, a thousand red streaks in the sundown sky, a thousand puffs up the trail. Everybody goes about it differently, of course, but I don't guess we'd trade any of it. It's meant a lot of good humor. It's meant a few flashes of almost unbearable beauty which I can only call religious experiences (and if religion means anything, that's what they were). "Fitness," experience, are part of it, too. Most important is an imperishable attitude, a philosophy if you like, a way of laying out the world and of planting ourselves in it. Now we know what is trivia and what is real.

The best of times and the worst of times. Now Science with its right hand unveils the more and more delicate machineries of life just before (or after) its left hand destroys them. The same ravaging giant who threatens to demolish it utterly on Earth is the only creature who can comprehend and glorify Creation... But no, we'd better not allow ourselves even that little egoism. Doesn't the crane whoop in celebration, the honker honk in celebration, the otter dive and slide in celebration, the coyote bark in celebration, the buffalo paw and grunt in celebration? We aim a black box and scratch on beaten wood pulp.

Actually, the eloquence of the wilderness is not a pattern for human eloquence. There lives no hardier fool than whoever shouts, "The scene inspired me to set pen to paper," or brush to canvas, or thumb to lyre. The wilderness inspires nothing but itself. Our babblings and scratchings resume in the den or studio, whenever things resume their comfortable and incorrect proportions.

Terry and Renny Russell, planet Earth, twentieth century after Christ. We live in a house that God built but that the former tenants remodelled—blew up, it looks like—before we arrived. Poking through the rubble in our odd hours, we've found the corners that were spared and have hidden in them as much as we could. Not to escape from but to escape *to*: not to forget but to remember. We've been learning to take care of ourselves in places where it really matters. The next step is to take care of the *places* that really matter. Crazy kids on the loose; but on the loose in the wilderness. That makes all the difference.

Terry Russell

Berkeley
February 6, 1965

But these are human things.
The point of it all is Out There, a little
beyond that last rise you can just
barely see, hazy and purple on the sky.
These pages are windows.
And windows are to see through.

I. TRIUMPH

...For afterwards a man finds pleasure in his pains,
when he has suffered long and wandered long. So
I will tell you what you ask and seek to know —HOMER

Dawn over the arid West.
It's Joshua Tree, Utah, central Nevada, Anza Borrego,
or somewhere off the highway near Barstow.
It's part of the geography of hope.

Man was born to wander.
Gypsies. The wandering Jew. Cabeza de Vaca and Jed Smith.
"We better get out and change the hubs."
"I'd like to see a tourist Chevrolet make *that*. We almost came apart."
"Think this sand will be murder on the way back?"

"Hey, stop! I think I see a mano."
"I haven't seen a tire track. Wonder
when anybody was here last."
"No, not up the wash, idiot, the road's
over there... or is it? Well, I'm damned,
I guess this is the end."
"We could probably make it farther if
we wanted to."

"Jeez, isn't this cool."

I'd rather wake up in the middle of nowhere than in any city on earth. —STEVE MCQUEEN

The car has made our cities uninhabitable. It is also the best way to escape them. Hurry and take the road to the roadless area, because it won't be roadless long. Too much demand.

The gas pump doesn't know the beauty which it helped to see; and so the gas tax comes pouring in and the pavement comes pouring out.

And so we push the Big Wheel nearer the edge. The land of the free and the home of the auto dump. But man was born to wander.

As I was leaving the Irishman's roof after the rain, bending my steps again to the pond, my haste to catch pickerel, wading in retired meadows, in sloughs and bog-holes, in forlorn and savage places, appeared for an instant trivial to me who had been sent to school and college; but as I ran down the hill toward the reddening west, with the rainbow over my shoulder, and some faint tinkling sounds borne to my ear through the cleansed air, from I know not what quarter, my Good Genius seemed to say...

Go fish and hunt far and wide day by day—farther and wider—and rest thee by many brooks and hearth-sides without misgiving...

Remember thy Creator in the days of thy youth. Rise free from care before the dawn, and seek adventures. Let the noon find thee by other lakes, and the night overtake thee everywhere at home. There are no larger fields than these, no worthier games than may here be played...

Grow wild according to thy nature, like these sedges and brakes, which will never become English hay. Let the thunder rumble; what if it threaten ruin to farmers' crops? that is not its errand to thee. Take shelter under the cloud, while they flee to carts and sheds. Let not to get a living be thy trade, but thy sport. Enjoy the land, but own it not...

Through want of enterprise and faith men are where they are, buying and selling, and spending their lives like serfs.

—THOREAU

One of the best-paying professions is *getting ahold* of pieces of country in your mind, learning their smell and their moods, sorting out the pieces of a view, deciding what grows there and there and why, how many steps that hill will take, where this creek winds and where it meets the other one below, what elevation timberline is now, whether you can walk this reef at low tide or have to climb around, which contour lines on a map mean better cliffs or mountains. This is the best kind of ownership, and the most permanent.

It feels good to say "I know the Sierra" or "I know Point Reyes." But of course you don't—what you know better is yourself, and Point Reyes and the Sierra have helped.

We leave: part of ourselves.
We take: sand in our cuffs, rocks, shells,
moss, acorns, driftwood, cones, pebbles,
flowers,
Photographs.

But is the picture a tenth of the thing?
A hundredth?
Is it anything without the smell and salt
breeze and the yellow warmth when
the fog lifts?

Oh! but I got all that, too.
It is exposed for ever on the sensitive
emulsion sheet
Of my mind.

It's a shame that a race so broadly
conceived should end with most lives
so narrowly confined.
Why should we waste
Childhood on the children,
Poverty on the poor,
Antiquity on the antiquarians,
Or woods on the woodsmen?

They sleep generally in the open air, in winter as well as in summer, subjected to every inclemency of the weather. As may well be imagined, a buffalo hunter, at the end of the season, is by no means prepossessing in his appearance, being, in addition to his filthy aspect, a paradise for hordes of nameless parasites. They are yet a rollicking set, and occasionally include men of intelligence, who formerly possessed an ordinary amount of refinement.

—J. A. ALLEN

So why do we do it?
What *good* is it?
Does it teach you anything?
Like determination? invention? impro-
visation?
Foresight? hindsight?
Love?
Art? music? religion?
Strength or patience or accuracy or
quickness or tolerance or
Which wood will burn and how long
is a day and how far is a mile
And how delicious is water and smoky
green pea soup?
And how to rely
On your
Self?

How far *is* a mile?
Well, you learn that right off.
It's peculiarly different from ten tenths on the odometer.
It's one thousand seven hundred and sixty steps on the dead level and if you don't have anything better to do you can count them.
"One and a half? You're crazy, Jere, we've been walking for *hours!*"

It's at least ten and maybe a million times that on the hills
And no river bed ever does run straight.

"What's this, Frog Creek?
Is that all the further we are?
Look, tomorrow we *gotta* start earlier."

Red exhaustion rips at your throat
And salt sweat spills off your forehead and mats
your eyelids and brows
And drips on the burning ground
And your legs start to turn to rubber and collapse
like a balloon.
"Pretty soon I've got to rest.
How much farther? What's the good of this God
damn work anyway?"

The long distance runner is paid by the snap of a
white thread across his chest.
You are paid by the picture at your feet.

You can feel the muscle knots tightening in your legs
And now and then you reach down to test the hard
lumpiness.
The passes get easier and finally you're just laughing
over them.
Every step and every strain and hard breath and heart-
pump is an investment in tomorrow morning's strength.
You're watching the change with your own eyes and feeling
it under your own skin and through your own veins.
Fibers multiply and valves enlarge and walls thicken.
A miracle.

At least if the species has lost its animal strength
Its individual members can have the fun of finding it
again.

No servant brought them meals: they got their meat out of the river, or went without. No traffic cop whistled them off the hidden rock in the next rapids. No friendly roof kept them dry when they mis-guessed whether or not to pitch the tent. No guide showed them which camping spots offered a nightlong breeze, and which a nightlong misery of mosquitoes; which firewood made clean coals, and which only smoke... The elemental simplicities of wilderness travel were thrills not only because of their novelty, but because they represented complete freedom to make mistakes. The wilderness gave them their first taste of those rewards and penalties for wise and foolish acts which every woodsman faces daily, but against which civilization has built a thousand buffers. These boys were "on their own" in this particular sense. —Aldo Leopold

Play for more than you can afford to lose, and you will learn the game.

—Churchill

"Are you all by yourself?" asked the man and
his wife as suddenly I crunched through
the spring snow past their house trailer.
"Are you all by yourself?" asked the gull.
"Are you all by yourself?" asked the stars.

If a man is all by himself on this miraculous
earth, a neighbor is no help.

He was alone. He was unheeded, happy, and near to the wild heart of life. He was alone and young and willful and wild-hearted, alone amidst a waste of wild air and brackish waters and the seaharvest of shells and tangle and veiled grey sunlight. —JAMES JOYCE

One generation passeth away, and another generation cometh: but the earth abideth forever...

The sun also ariseth, and the sun goeth down,
and hasteth to his place where he arose...

The wind goeth toward the south, and turneth about unto the north; it whirleth about continually, and the wind returneth again according to his circuits...

All the rivers run into the sea; yet the sea is not full; unto the place from whence the rivers come, thither they return again. — ECCLESIASTES

Nature might have made Sphinxes in her spare time
Or Mona Lisas with her left hand,
Blindfolded.

Instead she gave the grain of sand,
The polished river stone,
The Grand Canyon.

So you went to the Louvre:
What did you see?

After the first Artist
Only the copyist.

So Man is not what he appears.
I had been blind a thousand years.

Wisdom older than the seers,
Beauty much too deep for tears,
And holy silence bursts the ears.

Ssh. The music of the spheres.

Talking to men who but lately had kissed their wives goodnight and slept under storm-tight roofs, they must have had a look in their eyes and a way of standing. Their shirts and breeches of buckskin or elkskin had many patches sewed on with sinew, were worn thin between patches, were black from many campfires and greasy from many meals. They were threadbare and filthy, they smelled bad, and any Mandan had a lighter skin. They gulped rather than ate the tripes of buffalo. They had forgotten the use of chairs. Words and phrases, mostly obscene, of Nez Percé, Clatsop, Mandan, Chinook came naturally to their tongues when they asked what word from Kentucky and was John Bull still fighting the frogs.

And still, men who by guts and skill had mastered the farthest wilderness, they must have had a way of standing and a look in their eyes. While they scanned the faces of white men, their glance took in the movement of river and willows, of background and distance. While they talked as men talk nearing home and meeting someone newly come from there, their minds watched a scroll of forever-changing images. What they had done, what they had seen, heard, felt, feared—the places, the sounds, the colors, the cold, the darkness, the emptiness, the bleakness, the beauty. Till they died this stream of memory would set them apart, if imperceptibly to anyone but themselves, from everyone else. For they had crossed the continent and come back, the first of all.

—Bernard DeVoto

We, the last of all, are the first of all.
The oldest is always the newest,
I see nothing which I have seen before;
A man is never lost, he has only been mislaid.

Got to move on, got to travel, walk away my blues.

Well,
Have we guys learned our lesson?
You bet we have.
Have we learned to eschew irresponsible outdoorsmanship, to ask advice, to take care and to plan fastidiously and to stay on the trail and to camp only in designated campgrounds and to inquire locally and take enough clothes and keep off the grass?
You bet we haven't.
Unfastidious outdoorsmanship is the best kind.

Adventure is not in the guidebook
and Beauty is not on the map.

Seek and ye shall find.

Now I see the secret of making the best persons,
It is to grow in the open air and to eat and sleep
with the earth. —WALT WHITMAN

II. TRAGEDY

Man always kills the thing he loves, and so we pioneers have killed our wilderness. —Aldo Leopold

1963

"We hereby demand more *Jobs!*
We hereby demand more *Power!*
Let it go on record right here and now that we hereby pledge
to *Tame This Treacherous Torrent!*
[applause.]
We hereby demand more Recreation!
We hereby demand more Reclamation!
We hereby demand more ECONOMIC GROWTH!
We hereby demand more... PROGRESS!!"
[thunderous applause.]

Spin the wheels. Faster. *Hum whirl flash rumble hammer revolve explode.*
Grease the gears with outboard oil.
Grease the gears with the fat of beaver who aren't any use.
Grease the gears with the blood of deer who aren't any use.
Grease the gears with dissolving cottonwoods and the sickly sweet perfume they wear when they drown.
Grease the gears with the stale slime on the shore as the banks fall over and as the grass and the moss and the brush and willows and reeds and seeds and pods sink underwater.
Grease the gears with my and your blood and the blood of everyone who floated down and lost himself in the side canyons and on the riffles and sand bars
And left part of himself on the walls.
We're all under water now, and drowned.
We burst the ranks of the walking dead, and the killer goes Unscathed.

19963

"No, it wasn't always this nice.
Most always, yeah, but for a little while the water didn't flow."
He shook his antlers and went back to browsing.

A man will never know again.
Thinking we won, we were the only losers.

The awful tragedy keeps welling up in your
hollow throat and chest.
A friend dies, or a President is shot.
A river is throttled.
My mouth is stale and my heart is under
my soles, I glare at the ground and spit
bitterly for days.
Now Little Eden is 200 feet underwater.
Maybe the first diversion tunnel closed an
eternity ago, when the continents were all
underdeveloped areas and the Tree of Life
grew underutilized and the mother of
us all took the first bite
In another Eden.

We must look funny to Someone,
Tumbling through the universe locked in a death grip with
our tiny ball Earth and ripping her busily to pieces,
trailing a stinking film of gas and pieces of satellites
and mushroom and dust clouds.

Think of her new.
An unspoiled country lying open to the sun.
Think of oceans of beauty, instead of scattered puddles,
muddy and drying up.

What can make the heart ache more than a billboard?

Is man the only mourner of wilderness because he is the
only killer?
Could we create if we could not destroy?
Would we want knowledge without control? beauty
without rape?
Is pastoral man a half-man and love a fiction?

Do I have nothing to blame but the genes in my own body?

Was I just born too late?

No.

My salvation is that I was not born into the adolescence of my race.

Its beautiful childhood may be gone, but its manhood is *now.*

Evolution is aware of itself.

At the last hour of the planting season, the seeds of a universal sanity are sown.

I look at a redwood and don't see board feet.

I look at a river and don't see kilowatt hours.

I look at a lake and don't see an aqueduct.

I look at a marsh and don't see more rotting surplus wheat.

I look at a gorge and don't see a damsite.

I look at a meadow and don't see real estate.

I look at an egret and don't see an absurd feathery hat.

The early settlers cluck and shake their heads, but the earliest settlers are glad.

The weed will win in the end, of course.

Time is on our side, boys, time is on our side.

Thine alabaster towns will tumble, thine engines
rot into dust.
Man will break his date with the future,
No matter how long he wants to play outlaw, no
matter how long he wants to gallop through
town shooting like a madman and hooting
at the laws of nature's god.
It is not they that he has made obsolete, it is
himself.

This knowledge is called wisdom.

But in these plethoric times when there is too much coarse stuff for everybody and the struggle for life takes the form of competitive advertisement and the effort to fill your neighbor's eye, there is no urgent demand either for personal courage, sound nerves or stark beauty, we find ourselves by accident. Always before these times the bulk of the people did not overeat themselves, because they couldn't, whether they wanted to or not, and all but a very few were kept "fit" by unavoidable exercise and personal danger. Now, if only he pitch his standard low enough and keep free from pride, almost anyone can achieve a sort of excess. You can go through contemporary life fudging and evading, indulging and slacking, never really hungry nor frightened nor passionately stirred, your highest moment a mere sentimental orgasm, and your first real contact with primary and elemental necessities the sweat of your deathbed. —H.G. Wells

God secure me from security, now and forever, Amen.

Who's afraid of the universe?
It's midnight on the desert or the coast or high
above timberline, the Milky Way is close and
the stars are singing.
I am not small, I fill the sphere.
I tremble before the cosmos no more than a
fish trembles before the tides.

We fear what we don't know:
I know what the hills are there for and they
know me.

Cut the root and the plant dies.
City life is the scary life, inane, insane, tiny
and alone.
Learn wildness and you don't fear anything.
Except people afraid.

...And now the villagers, who scarcely know where it lies, instead of going to the pond to bathe or drink, are thinking to bring its water, which should be as sacred as the Ganges at least, to the village in a pipe, to wash their dishes with!—to earn their Walden by the turning of a cock or drawing of a plug!...Where is the country's champion, the Moore of Moore Hall, to meet the enemy at the Deep Cut and thrust an avenging lance between the ribs of the bloated pest?—THOREAU

So as long as I can see I will keep looking.
As long as I can walk I will keep moving.
As long as I can stand I will keep fighting.

At the end of All the Ages
A Knight sate on his steed,
His armour red and thin with rust,
His soul from sorrow freed;
And he lifted up his visor
From a face of skin and bone,
And his horse turned head and whinnied
As the twain stood there alone.

No bird above that steep of time
Sang of a livelong quest;
No wind breathed,
Rest:
"Lone for an end!" cried Knight to steed,
Loosed an eager rein—
Charged with his challenge into Space:
And quiet did quiet remain.

—Walter de la Mare

THE END

SOUTHERN CALIF.
MESQUITE SPRING 22 1/2
SAND SPRING
WATER

... and so there ain't nothing more to write about, and I am rotten glad of it, because if I'd a knowed what a trouble it was to make a book I wouldn't a tackled it and I ain't agoing to no more. But I reckon I got to light out for the Territory ahead of the rest, because Aunt Sally she's going to adopt me and sivilize me and I can't stand it. I been there before. — MARK TWAIN

PHOTOGRAPHS

*OTHER PHOTOGRAPHER

88 Foot of Rainbow Bridge, Rainbow Bridge National Monument, Utah (June 1964).
*88 Unknown (unknown date).
90 Lake Powell, Utah (formerly Aztec Creek), (June 1964).
92 Lake Powell, Utah (formerly Mystery Canyon), (June 1964).
*94 Colorado River at Driftwood Canyon, Utah (pre-1963). By Philip Hyde.
96 Lake Powell, Utah (formerly Colorado River, near Gunsight Canyon), (June, 1964).
98 Davis Gulch, off Escalante River, Utah (July 1964).
99 Gold Standard Mine, Joshua Tree National Monument, California (December 1964).
100 Lake Powell, Utah (formerly Colorado River, near Hole-in-the-Rock), (June 1964).
102-03 Cathedral in the Desert, Clear Creek, off Escalante River, Utah (June 1964).
104 Belmont, Nevada (November 1964).
106 Graveyard near San Quintin, Baja California (September 1964).
108 Pacific Coast off California State route 1 near Stinson Beach (March 1964).
110 Lake Lagunitas, Marin County, California (April 1964).
113 Lake Powell, Utah (formerly Colorado River, near Gunsight Canyon), (June 1964).
*114 Sea cliff near Stinson Beach, California (October, 1964). By Jerry Joiner.
117 South Beach, Point Reyes National Seashore, California (September 1964).
*118 Unknown (unknown date).
BACK COVER Near Wishon Dam, High Sierra, California (August 1962).

To describe *On the Loose* is to deprive the reader of a freshness that ought to be self-discovered, and turned to often and remembered. It is better to say something about the authors.

Renny Russell [Sumner], the younger of the brothers who conceived of and executed *On the Loose*, learned about wilderness early, especially about the Sierra Nevada, the Pacific shore, and the Plateau Province of Colorado and Utah. So did Terry, a brilliant student at the University of California, Berkeley, and a staunch advocate of free speech there and elsewhere, and principal architect of this book. On the trip through Glen Canyon I shared with them, they had been out so long that the Navajo sandstone was pale by comparison, and they had become as much a part of the country as it was. They were rafting down the Green River in June 1965 when, as they rounded a bend in the river, a rapid surprised them and the raft upset. Terry was lost.

When Terry brought the manuscript up to our house earlier that spring, its text, illustrations, calligraphy, and design beautifully complete. I knew the Sierra Club must somehow publish it, and would be proud of having and using the chance. Terry knew this too.

DAVID BROWER
Executive Director, Sierra Club

New York City, December 18, 1966